YOUR KNOWLEDGE HAS VALUE

- We will publish your bachelor's and
  master's thesis, essays and papers

- Your own eBook and book -
  sold worldwide in all relevant shops

- Earn money with each sale

Upload your text at www.GRIN.com
and publish for free

Bibliographic information published by the German National Library:

The German National Library lists this publication in the National Bibliography; detailed bibliographic data are available on the Internet at http://dnb.dnb.de .

Imprint:

Copyright © 2018 GRIN Verlag
Print and binding: Books on Demand GmbH, Norderstedt Germany
ISBN: 9783668749924

This book at GRIN:

https://www.grin.com/document/428648

Leonard Kahungu

Life Extension Of Oil And Gas Assets In The North Sea. Health and Safety Issues

GRIN Verlag

HEALTH AND SAFETY REGARDING AGEING AND LIFE EXTENSION OF OIL AND GAS ASSETS IN THE NORTH SEA

BSc (Hons) Oil and Gas Management

Table of Contents

Introduction

Royal Dutch Shell plc, simply referred to as Shell, is a British-Dutch multinational company dealing with oil and gas industry, which has its headquarters in the Netherlands and incorporated in the UK (Jansen et al., 2016). Shell has vast interests in the oil and gas industry, including refining transport, exploration and production, power generation, petrochemicals, trading, distribution and marketing, and recently renewable energy. The company operates in over 70 countries with over 44,000 service stations across the globe and produces more than 3.7 million oil barrels per day (Palombo, 2015). Consequently, the offshore Shell operations have been faced with considerable challenges in the North Sea, due to dwindling oil resources. Currently, major oil operations are targeting marginal fields and even by seeking more challenging fields like the West of Shetland and event going further into deeper sea oceans. Initially, Shell had closed much of its operations in the North Sea, until recently, 2018, where the company has relaunched its operations for the first time in more than six years (Samarakoon and Ratnayake, 2015). The Penguins fields, which are located in the north-east of the Shetland Islands and it is projected to involve the construction of floating production, storage and offloading vessel with the ability to reach a peak output of an average of 45,000 barrels of oil equal per day.

Established fields are characterized by reducing production rates and increasing unit expenses of the structures that have been functioning beyond the intended design lifespan. Besides, oil and gas production process integrates the functions of numerous pieces of equipment, which are subjected to rigorous operating environments. Corrosion, wear, erosion and ageing-related factors weaken the conditions of these facilities (Crompton, 2016). These factors can cause unexpected failures, and it leads large economic losses, and most significantly, health and safety issues to personnel and the living things within the proximity of the severe environmental pollution.

Ageing and Life Extension in the North Sea

The major concerns arising from ageing and life extension (ALE) present significant challenges for offshore oil and gas operations. Technological advancements in oil and gas exploitation and production indicate that the United Kingdom Continental Shelf (UKCS) has multiple establishments that are currently operating beyond their initially expected field life (Stacey, 2011). The ageing process of equipment increases the challenges in maintaining the integrity, which necessitates the integrity maintenance. Cumulative degradation process often triggers this to an extent where replacements are no longer sustainable due to significant changes in engineering standards. Besides, increased reliance on the existing infrastructures to operate as hubs for exploring prospectus fields create the need for life extension (Aeran et al., 2016). These factors explicitly indicate that the need for extending the use of existing facilities in offshore operations will continue to grow. This necessitates the significance of understanding the ageing and life extension processes and implement effective intervention measures on demand without jeopardizing the integrity, health, and safety of these facilities.

The ageing process is defined as the universal process in which the features of systems, structures, or components progressively depreciate with time and use. This approach addresses the durability technical aspects, but the ageing process may include the facilities becoming obsolete or old and outdated due to technological advancements (Ratnayake, 2012). The need for replacing outdated pieces of equipment may be associated with social and economic factors. Besides, the ageing process is viewed through the reliability-based, which addresses the gradual loss of operational capabilities, and physical ageing process that accounts for the gradual degradation of the equipment functions and properties (Stacey, 2011).

Risk Management and Asset Integrity

One of the most important components in asset ALE is the assessment of the changing conditions of the asset. This is accomplished through routine inspections, testing, and maintenance of the assets' integrity throughout their lifespan (Galbraith et al., 2005). As a consequence, systematic approaches supported by prompt repairs, routine replacement, and restoration of the asset's conditions are fundamental in the maintenance and life extension of oil and gas offshore assets. These activities ensure that the assets remain economically viable without jeopardizing the safety and critical performance throughout the life of these assets (Palombo, 2015). The failure to perform regular maintenance leads to integrity deterioration, which reduces the performance, speeds up the ageing process.

Nonetheless, the management of review, testing, and maintenance are often lifelong processes that target the asset depreciation mechanisms affecting plants and safety critical elements, along with their sub-systems and components (Nezamian and Altmann, 2013). Essentially, the process had to anticipate, review and intervene to degradation possibilities, including erosion, fatigue, corrosion and other mechanisms presenting asset ageing risks and addressed before the failure, or safety issue occurs. An asset register consists of comprehensive details of in-service components, in which the failure could threaten primary accident prevention objectives in the plant (Jansen et al., 2016). This can be integrated into systems specially designed for inspection, review, and maintenance management functions. The asset register integrates assets and components that are installed permanently.

New assets and components triggered by modification including defined life repairs and temporary pieces of equipment are integrated into the asset register. Thus, supplementary inspections, reviews, and maintenance activities should be aligned with the asset register (Crompton 2016). This shows the significance of putting in place periodic reviews in attempts to ensure that the asset register along with supplementary database is maintained and that they are of up-to-date. Besides, rigorous management of the ageing process will enhance the asset register, which is fundamental in extending the life of offshore plants (Wright, 2011).

Consequently, the integrity risk assessment entails the review of the asset register to establish the risks of failure in the asset's plant and equipment. This type of assessment is influenced by the type of components being reviewed and integrates failure modes, impacts and critical evaluation (Shafiee et al., 2016). Risk-based inspection reviews and other forms of suitable assessments that are critical to the primary functions of a component are also included in integrity risk assessment. The inspection, testing, and maintenance programme is another important component of asset ALE in offshore facilities (Christie et al., 2015). The primary factors, in this case, include function testing of safety-critical evaluation, risk-based inspection, reliability centralised maintenance, along with planned predictive and preventative maintenance.

Technology

The contemporary world is characterised by advanced technologies that have revolutionised nearly all industries. This has led to the emergence of high-tech applications, which play a central role in supplement various inefficiencies attributed to human weaknesses (Samarakoon and Ratnayake, 2015). However, these developments are presenting new and unique challenges, particularly in the oil and gas industry. Technological applications are associated with various advantages, especially by improving the efficiency of the production and exploitation processes. Organisations with vast interests in the North Sea offshore dealing increasingly appreciate the significance of knowledge and technology sharing (Leon et al., 2017). This has enhanced the ability to address emerging technical issues in the typical operations. As a consequence, emerging technology continues to enhance the efficiency of the ageing and life extension management processes.

Additionally, technical advancement compels organisations dealing with oil and gas industries to adhere to technical requirements that protect the health and safety concerns arising from the working conditions along with the environment (Hutchison et al., 2015). Even though technological advancements contribute positively to ALE management, they bring new challenges at the organisational and structural levels. The modern technology compels companies in this field to observe technical requirements. This increases the legal and regulation burdens, imposed by HSE and ALE guidelines.

ALE and Health and Safety Executive Guidelines

The Health and Safety Executive (HSE) is the body responsible for regulating oil and gas industry in the UK. HSE champions the reduction of risks attributed to the ageing assets in the UK. Earlier studies assessing the efficiency of HSE Key Programme shows that oil and gas industry could leverage on extending the lives of the plants, making it mandatory for oil and gas plants in the North Sea submit compliance reports regarding assets ALE management (Liu et al., 2015). Consequently, the Major Accident Reporting System (MARS) database covering entire Europe shows that the loss of containment is related to the ageing processes in the plan facilities. The MARS database was used to evaluate various incidents taking place between 1980 and 2006. This report indicates that 96 incidents reported were associated with the poor management of ALE process. The incident led to the loss of eleven lives and caused over one hundred and eighty-three injuries. These findings were used to conclude that more than 60% of the accidents recorded in the MARS database are linked to technical asset integrity in most ageing plant facilities (Little et al., 2016). It was also established that over 50% of the plants operating in the UK continental shelf had exceeded their intended lifespan of 25 years. As a consequence, these facilities were identified as a major health and safety concern, which led to the development of Key Programme 4 by the Health and Safety Executive.

The Key Programme 4 (KP4) compels organisations to submit risk management and asset integrity plans in the process of enhancing the quality and life of the oil and gas industry. It is paramount to understand ways in which HSE is aligned with the inspection and investigation guidelines aligned to KP4 programme and its significance (Devold et al., 2017). In other words, KP4 plays a central role in establishing the degree to evaluate the integrity risks arising from the ageing and life extension processes enacted by oil and gas industries. KP4 promotes the development of good practices in the offshore industry in attempts to enhance the safety of these facilities. Four areas of concern include the integrity, safety culture, competence and leadership (Sharp et al., 2015). Notably, the program is not primarily concerned with the physical age of offshore assets but focuses on improving the conditions and other factors that influence the commencement, progression, and alleviation of the degradation process.

The HSE aligns the KP4 inspection programme through a set of objectives to strengthen the effectiveness of the latter. KP4 objectives ensure that various vulnerabilities of the asset are sufficiently addressed by addressing integrity issues arising from the management of the ageing and life extension of the infrastructures (Sandana et al., 2016). Thus, one of the KP4 objectives is to increase awareness of the importance of effective management of ALE. The second objective is to inspect individual organisations and the ALE strategies enacted. KP4 also seeks to identify various challenges and implement appropriate programme solutions as manifested through the identification of health and safety shortcomings (Rosen et al., 2016). The last objective is concerned with the development of a common industrial approach in the management of ageing installations, in attempts to guarantee the health and safety, particularly in the operations of offshore installations placed on the UK continental shelf.

Economic Impacts

The primary aim of the ALE is often concerned with the profitability and health and safety issues concerning the structural integrity of the assets. In particular, ALE ensures that the assets to continue to function within the expected capacity range. This is mainly done on economically viable offshore plants since assets that have extended beyond their functioning abilities are often decommissioned (Aeran et al., 2016). ALE ensures that proceedings obtained from certain assets are obtained for longer periods. The resulting extension is relatively cheaper compared to the installation of new pieces or equipment or complete construction of a new plant, which requires an immense amount of financial and human resources to improve the structural and functional integrity (Stacey, 2011). Therefore, ALE promotes the efficiency of exploiting financial and human resources by extending the life of assets and hence reducing the need for new installations.

Poor ALE management increases the risk of health and safety concerns. As previously observed on the MARS database, it is evident that most accidents recorded on the offshore oil and gas fields result from ageing facilities and asset integrity deterioration (Ratnayake, 2012). The failure of upholding asset integrity escalates economic liabilities, interfering with financial difficulties. This could limit the exploitation and operation efficiencies attributed to increased liabilities. Thus, ALE plays a central role in reducing the cost of oil and gas exploration, exploitation, production, and management (Galbraith et al., 2005). It addresses the safety and health concerns resulting from the facilities' assets.

Technology

Nonetheless, ALE management is accompanied by additional management and installations costs. The management of ageing and life extension processes focus on the continuous or lifelong maintenance process that elicits the need for new replacement and intensified maintenance activities (Nezamian and Altmann, 2013). The cost of regular maintenance is known to be prohibitive and may increase the management expenses. Replacing obsolete equipment requires a significant amount of financial resources. This indicates technology is a significant economic determinant in ALE process (Palombo, 2015). The invention of new equipment with advanced efficiency levels necessitates the replacement process in attempts to increase the lifespan of the project.

Furthermore, technology plays a central role in the management of ALE process. It facilitates the inspection and registry management. This enhances the efficiency of the management process, enhancing the efficiency of evaluating the ageing process (Jansen et al., 2016). Computerised applications improve the efficiency of inventory keeping, which is critical in tracking the ageing components. This plays a central role in safeguarding the integrity of the offshore assets (Crompton 2016). The ageing and life extension management must accommodate high-tech application in protecting the structural integrity of these assets and prolonging the lifespan in offshore facilities by delaying the ageing process. Integrating life extension technologies with the health and safety aspects will further safeguard the working conditions of the human resources.

Environmental Issues

The main health and safety hazards are described as multiple accidents that could result in multiple fatalities and increase the spread of property damage or cause significant environmental issues. Environmental hazards in the recent offshore operations history have a low frequency, but the impacts are lethal (Wright, 2011). Oil spillages are also of great concerns in the offshore oil and gas operations. HSE recognises hydrocarbon releases as a major environmental, health and safety concern (Samarakoon and Ratnayake, 2015). As a consequence, there exist significant hydrocarbon release system guidelines which provide guidelines on the appropriate measures in the release of the hazards.

The lifting and movable units such as the drilling platforms in the offshore equipment are required to comply with the HSE guidelines regarding the environmental impact. Some of these requirements include low hydrocarbon release systems and null spillages in the North Sea operations (Christie et al., 2015). It is imperative to integrate the environmental factors in the management of ageing and life extension processes. Notably, ALE primary objectives are based on prolonging the economic viability of the ageing assets beyond their designed lifespan (Sandana et al., 2016). This makes offshore assets prone to numerous environmental conditions, which when poor managed can lead to catastrophic health and safety incidents.

Weather conditions in the North Sea are considered as harsh and increase the risk of fastening the ageing processes of the facilities found in this region (Rosen et al., 2016). These conditions are likely to hasten the ageing process, reducing the economic viability of linked to ALE management. Harsh weather conditions characterising oil and gas plant assets threatens the extension processes (Sharp et al., 2015). A high level of salinity increases the corrosion process, while the strong sea waves threaten the structural integrity. Thus, the ageing process must account for the weather conditions by guaranteeing the health and safety of the oil and gas assets.

A business case about ageing and life extension in the North Sea

Presently, oil and gas industries operating in the North Sea are striving to maximise the economic viability of offshore assets. This is attributed to the general observations that most assets in this region have exceeded their intended lifespan and are now seeking to maintain their productivity in the midst of stringent HSE and ALE guidelines in the region (Shafiee et al., 2016). Nonetheless, it is imperative to note that ALE is not only done in the ageing assets but have also become integral parts of the offshore operations. This is exemplified by the case of Shell, in which the organisation is seeking partnerships to improve the returns linked to certain plants and assets in the UK continental shelf (Devold et al., 2017). Nonetheless, these approaches in the exploitation of dwindling hydrocarbon reserves have, however, contributed to emerging challenges and regulations seeking to guarantee the integrity and safety of the offshore plant assets.

Consequently, the ALE regulations provide fundamental legal and structural guidance that ensures that the extension of these facilities does not jeopardize the structural integrity, health, and safety of the workers and the environment at large (Little et al., 2016). HSE as provided by the ALE guidelines and the KP4 objectives targets individual components in the assets installed in the offshore plants (May et al., 2008). Thus, major and minor installations or modifications are laid in accordance with the guidelines provided by the UK and European Union guidelines (Leon et al., 2017). This is attributed to the fact that modifications necessities the implementation of change management processes as laid out by the existing guidelines. Extending the lifespan of offshore oil and gas assets means pushing the limits of technologic and engineering theoretical concepts, which threaten the asset integrity and the working environments.

Conclusion

Evidently, management approaches applied in the ageing and life extension processes are fundamental in prolonging the life and economic viability in oil and gas facilities, particularly those that are located in offshore zones. Shell, among others companies operating in the North Sea region, are increasingly focusing on the ageing and life extension management processes. The ageing and life extension processes refers to an all-purpose process that reduced the asset depreciation process, with a particular focus on corrosion, reduced functioning abilities, and replacing obsolete components. Although the ALE plays a central role in improving the structural integrity, it triggers serious health and safety concerns to the environment and the human resources managing oil and gas production plants. Primarily, ALE prolongs the life of the assets, which jeopardizes the structural integrity of the offshore plants. Besides, hydrocarbon reserves distributed across the North Sea region are prone to harsh environmental factors such as strong waves and high salinity levels. This is likely to speed up the ageing process, which also reduces the structural integrity of the facilities found in the North Sea. These factors illustrate the magnitude of health and safety concerns arising from ALE management. ALE necessitates the implementation of rigid health and safety guidelines as provided by the HSE procedures and other relevant regulations.

References

Aeran, A., Siriwardane, S.C. and Mikkelsen, O., 2016, June. Life Extension of Ageing Offshore Structures: Time dependent corrosion degradation and health monitoring. In *The 26th International Ocean and Polar Engineering Conference.* International Society of Offshore and Polar Engineers.

Christie, R., Fleming, C., Donnelly, B. and Huff, S., 2015, September. How Visualization Technology is Maximising Up Time. In *SPE Middle East Intelligent Oil and Gas Conference and Exhibition.* Society of Petroleum Engineers.

Crompton, J., 2016, September. How can we turn intelligent energy into profitable operations? In *SPE Intelligent Energy International Conference and Exhibition.* Society of Petroleum Engineers.

Devold, H., Graven, T. and Halvorsrød, S.O., 2017, May. Digitalization of Oil and Gas Facilities Reduce Cost and Improve Maintenance Operations. In *Offshore Technology Conference.* Offshore Technology Conference.

Galbraith, D.N., Sharp, J.V. and Terry, E., 2005, January. Managing life extension in ageing offshore Installations. In *Offshore Europe.* Society of Petroleum Engineers.

Hutchison, Emily, John Wintle, Alison O'Connor, Emilie Buennagel, and Clement Buhr. "Fatigue Reassessment to Modern Standards for Life Extension of Offshore Pressure Vessels." In *ASME 2015 Pressure Vessels and Piping Conference*, pp. V007T07A018-V007T07A018. American Society of Mechanical Engineers, 2015.

Jansen, H.M., Van Den Burg, S., Bolman, B., Jak, R.G., Kamermans, P., Poelman, M. and Stuiver, M., 2016. The feasibility of offshore aquaculture and its potential for multi-use in the North Sea. *Aquaculture International, 24*(3), pp.735-756.

Leon, M., Ahiaga-Dagbui, D.D., Fleming, C. and Laing, R.A., 2017. Exploring visual asset management collaboration: learning from the oil and gas sector.

Liu, H., Shi, X., Chen, X. and Liu, Y., 2015. Management of life extension for topsides process system of offshore platforms in Chinese Bohai Bay. *Journal of Loss Prevention in the Process Industries, 35*, pp.357-365.

Little, J., Fleming, C. and Donnelly, B., 2016, September. Using a forensic tool to save bed space with visual planning. In *SPE Intelligent Energy International Conference and Exhibition.* Society of Petroleum Engineers.

May, P., Sanderson, D., Sharp, J.V. and Stacey, A., 2008, January. Structural integrity monitoring: Review and appraisal of current technologies for offshore applications. In

ASME 2008 27th International Conference on Offshore Mechanics and Arctic Engineering (pp. 247-263). American Society of Mechanical Engineers.

Nezamian, A. and Altmann, J., 2013, June. An oil field structural integrity assessment for re-qualification and life extension. In ASME 2013 32nd International Conference on Ocean, Offshore and Arctic Engineering (pp. V02BT02A014-V02BT02A014). American Society of Mechanical Engineers.

Palombo, D., 2015. Chandler v. Cape: An alternative to piercing the corporate veil beyond Kiobel v. Royal Dutch Shell. Brit. J. Am. Legal Stud., 4, p.453.

Ratnayake, R.C., 2012, July. Challenges in inspection planning for maintenance of static mechanical equipment on ageing oil and gas production plants: The state of the art. In ASME 2012 31st International Conference on Ocean, Offshore and Arctic Engineering (pp. 91-103). American Society of Mechanical Engineers.

Rosen, J., Potts, A., Sincock, P., Carra, C., Kilner, A., Kriznic, P. and Gumley, J., 2016, May. Novel Methods for Asset Integrity Management in a Low Oil-Price Environment. In Offshore Technology Conference. Offshore Technology Conference.

Sandana, D., Soltis, J. and Wilkinson, J., 2016, June. Effective Implementation of Asset Intergrity Management Systems with Visualisation Tools. In CORROSION 2016. NACE International.

Samarakoon, S.M. and Ratnayake, R.C., 2015. Strengthening, modification and repair techniques' prioritization for structural integrity control of ageing offshore structures. Reliability Engineering & System Safety, 135, pp.15-26.

Shafiee, M., Animah, I. and Simms, N., 2016. Development of a techno-economic framework for life extension decision making of safety critical installations. Journal of Loss Prevention in the Process Industries, 44, pp.299-310.

Sharp, J., Ersdal, G. and Galbraith, D., 2015, September. Meaningful and Leading Structural Integrity KPIs. In SPE Offshore Europe Conference and Exhibition. Society of Petroleum Engineers.Stacey, A., 2011, January. KP4: Ageing & Life Extension Inspection Programme-The First Year. In Offshore Europe. Society of Petroleum Engineers.

Wright, I., 2011, January. Ageing and life extension of offshore oil and gas installations. In Offshore Europe. Society of Petroleum Engineers.